# 3 South Carolina SC READY Grade 3 Math Practice Tests

*Full-Length Test Prep with Detailed Answer Explanations*

**Dr. A. Nazari**

# 3 Practice Tests to Get You Started!

Hey there, future math whiz! ⭐

This book has **3 full practice tests** to help you warm up for the real thing. Think of it like stretching before a big game — these tests will get your brain ready and show you what to expect!

👍 Three tests is the **perfect start**!

👍 Each one helps you feel **more ready**!

👍 You'll be surprised how much you **already know**!

Sharpen your pencil and let's get warmed up! 🔥

> 66 Three practice tests is a great way to start. Take your time with each one, and you'll feel more confident every step of the way! 99

# How to Use This Book

*Your quick-start guide to 3 great practice tests!*

## What's Inside This Book

- **3 Full-Length Practice Tests** — Each one covers all the Grade 3 math topics you need to know!
- **Answer Key with Explanations** — Find out why each answer is correct, not just what the answer is.
- **Reference Pages** — A math symbols chart and multiplication table you can peek at any time.
- **A Test Tracker** — Write down your scores and watch your confidence grow!

## A Simple 3-Test Plan

With just 3 tests, here's a great way to use them:

- **Test 1** — **The Warm-Up.** Take this test without a timer. Get comfortable with the question types. Don't worry about your score — just do your best!
- **Test 2** — **The Practice Round.** After reviewing Test 1, try this one with a timer (ask a grown-up!). Focus on the topics that were tricky last time.
- **Test 3** — **The Real Deal.** Treat this like the actual test: quiet room, timed, no peeking at answers. See how much you've improved!

## Multiple Choice

Pick the **one best answer** from choices A, B, C, or D. Not sure? Cross out the ones you know are wrong, then pick from what's left. That's a smart move!

## Short Answer

Write your answer **and** show your work! Even if your final answer isn't right, showing your steps can earn you credit. Use scratch paper if you need more room.

## 66 After Each Test 99

Flip to the Answer Key and check your work. For every question you got wrong, **read the explanation carefully**. Then write the tricky topics on your Test Tracker page. If you need extra help, grab our **Grade 3 Math Study Guide!**

Find more at
ViewMath.com/SC-Grade3

# 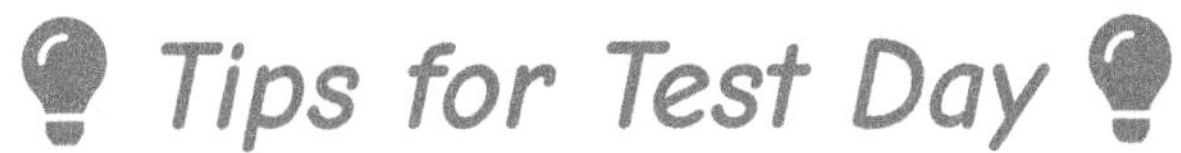 Tips for Test Day



*Easy tricks to help you feel calm and do your best!*

## 🌙 The Night Before

- ✅ **Sleep early** — your brain learns while you sleep!
- ✅ **Pack your supplies** — pencils, eraser, scratch paper, all ready to go.
- ✅ **Tell yourself:** *"I've been practicing. I'm going to do great!"*

## 👆 5 Simple Rules for Every Test

1. **Read the question twice.** The first time to understand it. The second time to catch details.
2. **Show your work.** Write the steps down, even on scratch paper. It helps you think!
3. **Skip the hard ones.** Put a small star next to tricky questions and come back later. Answer the easy ones first!
4. **Never leave a blank.** For multiple choice, your best guess is better than no answer at all.
5. **Check your work.** Finished early? Go back and re-read your answers.

## ✅ Smart Moves

- Take a deep breath before you begin
- Underline key words in the question
- Use drawings or number lines to help
- Cross out wrong answers first
- Double-check addition and subtraction

## ❌ Traps to Avoid

- Rushing and not reading carefully
- Picking the first answer that "looks right"
- Forgetting to carry or borrow numbers
- Skipping a question permanently
- Panicking when you see a tough problem

> **"** Remember, the very first practice test is the hardest — not because the questions are harder, but because everything is new! By Test 3, you'll feel like a pro. Trust me! **"**

# Get Ready to Practice

*Here's everything you need before you start!*

**Pencils**

*Sharpened and ready!*

**Eraser**

*Everyone makes mistakes!*

**Scratch Paper**

*For working things out*

**A Calm Spot**

*Somewhere quiet to focus*

**A Grown-Up**

*To help set a timer*

**A Can-Do Attitude**

*You've totally got this!*

## ✓ Allowed During Tests

- Pencils and erasers
- Blank scratch paper
- The **reference pages** in this book
- A ruler (for measurement questions)

## ✗ Not Allowed

- Calculators
- Phones, tablets, or computers
- Help from anyone else
- Your study guide (save it for after!)

## 👥 For Parents & Teachers

- With only 3 tests, **space them at least a week apart**. This gives time to review mistakes before trying the next one.
- Let your child take Test 1 untimed to build familiarity.
- After each test, go through the Answer Key together. Focus on **understanding the "why,"** not just the score.
- If a topic keeps tripping them up, review it in our **Grade 3 Math Study Guide** before the next practice test.
- Celebrate every bit of progress — even getting one more question right is a win!

# $X^1$ Math Reference Sheet $X^1$

*You may use this page during your practice tests!*

| Symbol | Name | What It Means | |
|---|---|---|---|
| $+$ | Plus (Add) | Put numbers together. | $3 + 5 = 8$ |
| $-$ | Minus (Subtract) | Take away from a number. | $9 - 4 = 5$ |
| $\times$ | Times (Multiply) | Add equal groups. | $4 \times 3 = 12$ |
| $\div$ | Divide | Split into equal groups. | $12 \div 3 = 4$ |
| $=$ | Equals | Both sides are the same. | $2 + 3 = 5$ |
| $>$ | Greater Than | The left number is bigger. | $7 > 3$ |
| $<$ | Less Than | The left number is smaller. | $2 < 9$ |
| $\frac{1}{2}$ | Fraction Bar | Part of a whole. | $\frac{1}{2}$ means 1 out of 2 equal parts |

## Key Math Words

- **Sum** — the answer when you add
- **Difference** — the answer when you subtract
- **Product** — the answer when you multiply
- **Quotient** — the answer when you divide
- **Factor** — a number you multiply
- **Array** — objects in rows and columns
- **Fraction** — a part of a whole
- **Numerator** — the top number in a fraction
- **Denominator** — the bottom number
- **Equation** — a math sentence with $=$
- **Estimate** — a smart guess, close to the real answer
- **Perimeter** — the distance around a shape
- **Area** — the space inside a shape
- **Rounding** — making a number simpler by going to the nearest ten or hundred

## 🔍 Word Problem Clue Words

- **Add** (+): in all, total, altogether, combined, sum, both, more
- **Subtract** (−): how many more, how many left, fewer, difference, remain
- **Multiply** (×): each, every, groups of, times, rows of, per
- **Divide** (÷): share equally, split, each group, how many groups, per

Find more at
ViewMath.com/SC-Grade3

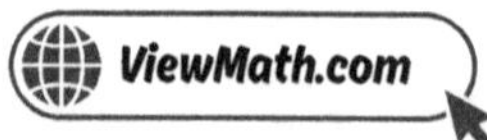

# ◫ Multiplication Table ◫



| × | 1 | 2 | 3 | 4 | 5 | 6 | 7 | 8 | 9 | 10 | 11 |
|---|---|---|---|---|---|---|---|---|---|----|----|
| 1 | 1 | 2 | 3 | 4 | 5 | 6 | 7 | 8 | 9 | 10 | 11 |
| 2 | 2 | 4 | 6 | 8 | 10 | 12 | 14 | 16 | 18 | 20 | 22 |
| 3 | 3 | 6 | 9 | 12 | 15 | 18 | 21 | 24 | 27 | 30 | 33 |
| 4 | 4 | 8 | 12 | 16 | 20 | 24 | 28 | 32 | 36 | 40 | 44 |
| 5 | 5 | 10 | 15 | 20 | 25 | 30 | 35 | 40 | 45 | 50 | 55 |
| 6 | 6 | 12 | 18 | 24 | 30 | 36 | 42 | 48 | 54 | 60 | 66 |
| 7 | 7 | 14 | 21 | 28 | 35 | 42 | 49 | 56 | 63 | 70 | 77 |
| 8 | 8 | 16 | 24 | 32 | 40 | 48 | 56 | 64 | 72 | 80 | 88 |
| 9 | 9 | 18 | 27 | 36 | 45 | 54 | 63 | 72 | 81 | 90 | 99 |
| 10 | 10 | 20 | 30 | 40 | 50 | 60 | 70 | 80 | 90 | 100 | 110 |
| 11 | 11 | 22 | 33 | 44 | 55 | 66 | 77 | 88 | 99 | 110 | 121 |

## ◉ How to Use This Table

To find **4 × 7**:

1. Find **4** in the left column (blue).
2. Find **7** in the top row (blue).
3. Follow the row and column until they meet: the answer is **28**!

# 📈 My Confidence Tracker 📈

*Record your scores below. You'll be amazed at your progress!*

**My name:** ___________________________________

| ✅ Test | 📅 Date | ⭐ Score | ☺ How I Feel |
| --- | --- | --- | --- |
| 1 | | | |
| 2 | | | |
| 3 | | | |



**The easiest topic for me was:**

___________________________________________________________

**The trickiest topic for me was:**

___________________________________________________________

**One thing I got better at from Test 1 to Test 3:**

___________________________________________________________

**Next time I want to try:**

___________________________________________________________

*You just finished 3 practice tests — that's awesome! Compare your first score to your last. I bet you'll see real improvement. Ready for more? Check out our 5-test or 7-test books for even more practice!*

Find more at
ViewMath.com/SC-Grade3

# Continue Learning at
# ViewMath Academy!

## For Parents, Teachers & Students

Great job on the practice tests! Want to keep improving? ViewMath Academy is your **free online companion** to this book.

- **Score Analyzer** — Enter your answers and instantly see which topics need more practice

- **Interactive Lessons** — Review the concepts behind each question with clear explanations

- **Adaptive Quizzes** — Practice your weak topics with questions that match your level

- **Progress Tracking** — See your mastery grow across all Grade 3 math topics

- **Personalized Dashboard** — A learning plan tailored just for you

**Scan to visit ViewMath Academy**

*viewmath.com/academy*

 Free to use · No downloads required · Works on any device

#  Table of Contents

*Here's what we'll explore together!*

 *Let's learn and have fun!*

# Practice Test 1

📋 30 Questions

## ✏️ Before You Start ✏️

- ✔ **Read each question carefully** before choosing your answer.
- ✔ **Show your work** on scratch paper when you need to.
- ✔ **Skip hard questions** and come back to them later.
- ✔ **Check your answers** when you're done.
- ✔ **Take your time** — there's no rush!

⭐ You've Got This! ⭐

Do your best and show what you know!

1. Which number has a 4 in the hundreds place and a 7 in the thousands place?

   (A)  4,731

   (B)  7,431

   (C)  7,341

   (D)  3,470

2. Which number rounds to 500 when rounded to the nearest 100?

   (A)  428

   (B)  449

   (C)  462

   (D)  551

3. A store sold 189 apples on Monday and 267 apples on Tuesday. How many apples were sold in all?

   (A)  356

   (B)  446

   (C)  456

   (D)  466

4. Use the standard algorithm to find $745 - 368$.

   Your Answer:

5. What is $6,005 - 2,738$?

   (A)  3,367

   (B)  3,267

   (C)  3,277

   (D)  4,267

6. Estimate $248 + 175 + 362$ by rounding each number to the nearest hundred.

   Your Answer:

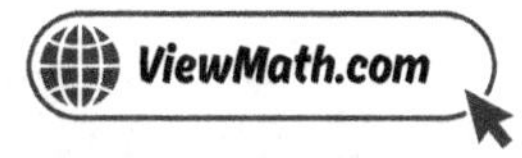

7. A spider has 8 legs. How many legs do 6 spiders have altogether? Write a multiplication sentence and solve.

*Your Answer*

8. A store has 5 shelves with 9 books on each shelf. How many books are there in all?

*Your Answer*

9. What is $8 \div 8$?

*Your Answer*

10. Which division fact is part of the same fact family as $9 \times 12 = 108$?

   (A) $108 \div 8 = 12$            (B) $108 \div 12 = 9$

   (C) $108 \div 11 = 9$            (D) $108 \div 9 = 11$

11. A bakery makes 5 trays of muffins with 8 muffins each. Then they make 12 more muffins. How many muffins in all?

   (A) 25            (B) 40

   (C) 52            (D) 60

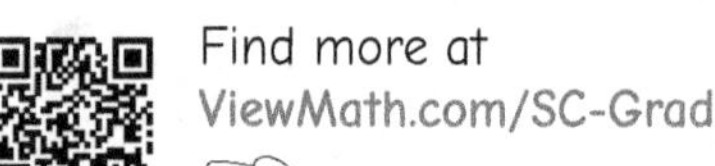

12. *Look at this table:*

| Input | Output |
|-------|--------|
| 12 | 4 |
| 18 | 6 |
| ? | 8 |
| 30 | 10 |

What is the missing input?

(A) 20

(B) 22

(C) 24

(D) 26

13. *What is $12 \times 4$?*

Your Answer:

14. *Look at the table. What is the missing input?*

| Input | Output |
|-------|--------|
| 3 | 18 |
| 5 | 30 |
| ? | 54 |

(A) 7

(B) 8

(C) 9

(D) 10

15. *A circle is split into 3 equal parts. None of the parts are shaded. Write the fraction that is shaded.*

Your Answer:

Find more at
ViewMath.com/SC-Grade3

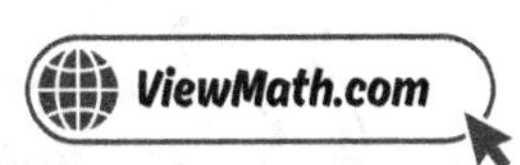

16. Would you rather have $\frac{1}{2}$ of a pizza or $\frac{1}{6}$ of the same pizza?

(A) $\frac{1}{6}$ because $6 > 2$

(B) $\frac{1}{2}$ because halves are bigger than sixths

(C) They are the same

(D) It depends on the pizza

17. Count by thirds: $\frac{1}{3}, \frac{2}{3}, ?$. What comes next?

(A) $\frac{3}{6}$

(B) $\frac{3}{3}$

(C) $\frac{4}{3}$

(D) $\frac{3}{9}$

18. Write 1 as a fraction with denominator 8.

Your Answer:

19. Compare: $\frac{3}{4} \underline{\phantom{=}} \frac{3}{8}$

(A) $>$

(B) $<$

(C) $=$

(D) Cannot compare

20. A string is $6\frac{1}{4}$ inches long. A ribbon is $6\frac{3}{4}$ inches long. How much longer is the ribbon than the string?

(A) $\frac{1}{4}$ inch

(B) $\frac{1}{2}$ inch

(C) $\frac{3}{4}$ inch

(D) 1 inch

21. A bag of apples has a mass of 3 kg. A bag of oranges has a mass of 5 kg. What is the total mass?

(A) 2 kg

(B) 15 kg

(C) 8 kg

(D) 35 kg

22. A large milk jug holds 4 liters. You pour out 1 liter for cereal. How much milk is left?

Your Answer:

23. How much is a quarter worth?

(A) 1 cent

(B) 5 cents

(C) 10 cents

(D) 25 cents

24. You buy a book for $4.50 and pay with a $10 bill. How much change do you get?

(A) $4.50

(B) $5.50

(C) $6.50

(D) $14.50

25. A bar graph has a scale that counts by 5s. A bar reaches up to the 4th line on the scale. What value does the bar show?

(A) 4

(B) 9

(C) 15

(D) 20

26. A line plot shows the lengths of 10 crayons. The data is shown below:

| 2 in. | $2\frac{1}{2}$ in. | 3 in. | $3\frac{1}{2}$ in. | 4 in. |
|---|---|---|---|---|
| 2 | 3 | 4 | 1 | 0 |

How many crayons are shorter than 3 inches?

(A) 2

(B) 4

(C) 5

(D) 9

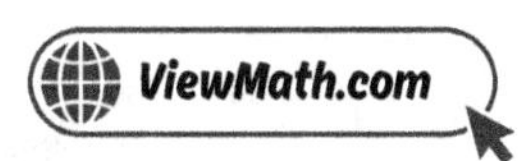

27. A shape has 4 sides. Two sides are 8 cm long and two sides are 3 cm long. It has 4 right angles. What shape is it?

Your Answer:

28. A square has sides that are 7 inches long. What is its area?

(A) 14 sq in

(B) 28 sq in

(C) 42 sq in

(D) 49 sq in

29. Lily says a rectangle that is 5 cm long and 3 cm wide has a perimeter of 15 cm. What mistake did she make?

(A) She subtracted instead of adding.

(B) She found the area instead of the perimeter.

(C) She forgot to add all 4 sides.

(D) She divided instead of multiplying.

30. If one pizza is cut into 4 equal slices and another identical pizza is cut into 8 equal slices, which slices are bigger?

(A) The $\frac{1}{8}$ slices

(B) The $\frac{1}{4}$ slices

(C) They are the same size.

(D) You cannot tell.

Find more at
ViewMath.com/SC-Grade3

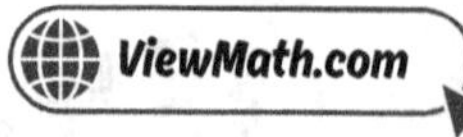

#  End of Practice Test 1 

Great job finishing the test!

##  My Score

I got ___________ out of 30 questions right.

Check your answers in the **Answer Key** at the back of the book.

💡 Review any questions you missed. That's how we learn!

## 📊 Check Your Score Online!

Visit **ViewMath Academy** to enter your answers and see which topics you need to review. You can also explore lessons, take quizzes, track your scores, and save your progress!

viewmath.com/score/3.1.SC.01

Or go to viewmath.com/score and enter code: 3.1.SC.01

2

# Practice Test 2

📋 30 Questions

## ✏️ Before You Start ✏️

- ✔ **Read each question carefully** before choosing your answer.
- ✔ **Show your work** on scratch paper when you need to.
- ✔ **Skip hard questions** and come back to them later.
- ✔ **Check your answers** when you're done.
- ✔ **Take your time** — there's no rush!

 ⭐ You've Got This! ⭐

Do your best and show what you know!

1. A number has 3 thousands, 7 hundreds, 0 tens, and 5 ones. What is the number?

Your Answer:

2. Maria rounded 438 to the nearest 10 and got 430. Is she correct?

(A) No, it should be 440

(B) Yes, 438 rounds to 430

(C) No, it should be 400

(D) No, it should be 450

3. What is $430 + 570$?

(A) 900

(B) 990

(C) 1,000

(D) 1,100

4. A farmer had 635 apples. He sold 278. How many apples does he have left?

Your Answer:

5. Use the standard algorithm to find $5,000 - 2,346$.

Your Answer:

6. Estimate $358 + 246$ by rounding each number to the nearest hundred.

Your Answer:

Find more at
ViewMath.com/SC-Grade3

7. Which skip counting sequence helps you find $4 \times 3$?

   (A) $3, 6, 9, 12$      (B) $4, 8, 12$

   (C) $3, 4, 5, 6$      (D) $1, 2, 3, 4$

8. A class has 5 rows of desks with 6 desks in each row. How many desks are there?

   (A) 11      (B) 25

   (C) 30      (D) 35

9. What is $20 \div 4$?

   (A) 4      (B) 5

   (C) 16      (D) 24

10. There are 55 crayons shared equally among 11 students. How many crayons does each student get?

   (A) 4      (B) 5

   (C) 6      (D) 11

11. Noah has $40. He buys 3 books that cost $7 each. How much money does he have left?

   (A) $12      (B) $19

   (C) $21      (D) $33

12. Which number is both a multiple of 4 AND a multiple of 6?

   (A) 8      (B) 12

   (C) 18      (D) 16

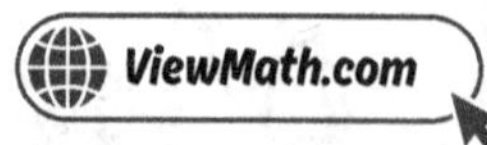

13. *What is $77 \div 11$?*

(A) 6                            (B) 7

(C) 8                            (D) 11

14. *Look at the table. What is the missing input?*

| Input | Output |
|-------|--------|
| 2     | 16     |
| 4     | 32     |
| ?     | 72     |

(A) 7                            (B) 8

(C) 9                            (D) 10

15. *Which fraction has a numerator of 2 and a denominator of 6?*

(A) $\frac{6}{2}$                          (B) $\frac{2}{6}$

(C) $\frac{2}{2}$                          (D) $\frac{6}{6}$

16. *What is special about ALL unit fractions?*

(A) *They all have the same denominator*      (B) *They all equal 1*

(C) *They all have 1 as the numerator*       (D) *They are all less than $\frac{1}{2}$*

17. *$\frac{7}{8}$ is built from how many copies of $\frac{1}{8}$?*

(A) 1                            (B) 7

(C) 8                            (D) 15

Find more at
ViewMath.com/SC-Grade3

18. *Write 4 as a fraction with denominator 8.*

Your Answer:

19. *Write* $<$, $>$, *or* $=$: $\quad \frac{1}{4} \underline{\phantom{=}} \frac{1}{8}$

Your Answer:

20. *A marker reaches from 0 to the half-inch mark past 6 inches. How long is the marker?*

Your Answer:

21. *A melon has a mass of 2 kg. A bunch of grapes has a mass of 500 g. How much heavier is the melon than the grapes?*

- (A)  498 g
- (B)  1,000 g
- (C)  1,500 g
- (D)  2,500 g

22. *Container A holds 20 liters. Container B holds 13 liters. How much more does Container A hold?*

- (A)  7 L
- (B)  13 L
- (C)  20 L
- (D)  33 L

23. *Liam buys an apple for $0.85 and a banana for $0.40. He pays with a $5 bill. How much change does he get?*

- (A)  $3.25
- (B)  $3.75
- (C)  $4.15
- (D)  $4.60

24. *Subtract: $12.00 − $7.45*

Your Answer:

25. Using the same rainy days graph (Jan = 6, Feb = 4, Mar = 3, Apr = 7), how many fewer rainy days were in February than in April?

(A) 1

(B) 2

(C) 3

(D) 4

26. Leo measured 8 worms. His line plot shows 7 X marks total. What went wrong?

(A) He measured one worm twice

(B) He forgot to plot one measurement

(C) He used the wrong numbers on the number line

(D) Nothing — 7 is correct

27. A polygon has 6 sides. What is the name of this polygon?

Your Answer:

28. A square has sides of 9 inches. What is the area?

Your Answer:

29. A triangle has sides of 4 cm, 6 cm, and 8 cm. What is the perimeter?

(A) 14 cm

(B) 18 cm

(C) 20 cm

(D) 24 cm

Find more at
ViewMath.com/SC-Grade3

30. *You fold a piece of paper in half, then fold it in half again. How many equal parts do you have?*

Your Answer:

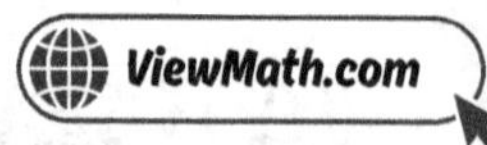

# 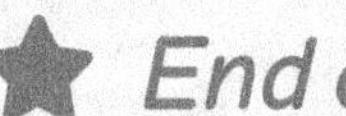 *End of Practice Test 2* 

*Great job finishing the test!*

## 📋 *My Score*

*I got __________ out of 30 questions right.*

*Check your answers in the **Answer Key** at the back of the book.*

💡 *Review any questions you missed. That's how we learn!*

## 📊 *Check Your Score Online!*

*Visit **ViewMath Academy** to enter your answers and see which topics you need to review. You can also explore lessons, take quizzes, track your scores, and save your progress!*

*viewmath.com/score/3.1.SC.02*

*Or go to viewmath.com/score and enter code: 3.1.SC.02*

# Practice Test 3

30 Questions

---

## ✏️ Before You Start ✏️

- ✔ **Read each question carefully** before choosing your answer.
- ✔ **Show your work** on scratch paper when you need to.
- ✔ **Skip hard questions** and come back to them later.
- ✔ **Check your answers** when you're done.
- ✔ **Take your time** — there's no rush!

★ You've Got This! ★

Do your best and show what you know!

1. Which number is the same as 9 thousands, 0 hundreds, 2 tens, and 5 ones?

(A) 9,250

(B) 9,025

(C) 9,205

(D) 9,520

2. When you round 549 to the nearest 10 and to the nearest 100, which gives the larger answer?

(A) Rounded to the nearest 10

(B) Rounded to the nearest 100

(C) They are the same

(D) Cannot tell without rounding

3. When using the standard algorithm, what do you do when a column adds to 10 or more?

(A) Skip that column.

(B) Start over from the beginning.

(C) Write the ones digit below the line and carry the tens digit to the next column.

(D) Subtract 10 from both numbers.

4. In the standard algorithm, what do you do when the top digit is smaller than the bottom digit?

(A) Switch the two numbers.

(B) Write zero and move on.

(C) Borrow 1 from the column to the left.

(D) Skip that column.

5. What is $7{,}000 - 3{,}254$?

(A) 3,746

(B) 3,856

(C) 4,746

(D) 3,756

6. Which rounding method gives a closer estimate for $463 + 219$?

(A) Rounding to the nearest 10

(B) Rounding to the nearest 100

(C) Both give the same estimate

(D) You cannot estimate this sum

7. How many stars are in 5 groups of 6?

(A) 11

(B) 25

(C) 30

(D) 56

8. What is $2 \times 9$?

(A) 11

(B) 16

(C) 18

(D) 20

9. What is $9 \div 1$?

(A) 0

(B) 1

(C) 9

(D) 10

10. What is $84 \div 12$? Explain how you used multiplication to find the answer.

Your Answer

11. Mom bakes 36 cookies and divides them equally onto 4 plates. Dad adds 3 cookies to each plate. How many cookies are on each plate now?

(A) 10

(B) 11

(C) 12

(D) 13

12. *What is the next number in this pattern?* $7, 14, 21, 28, \ldots$

(A) 30

(B) 32

(C) 35

(D) 36

13. *What is* $11 \times 12$*?*

(A) 112

(B) 121

(C) 132

(D) 144

14. *The rule is "Multiply by 8." Which pair of input and output is CORRECT?*

(A) *Input* = 5, *Output* = 13

(B) *Input* = 5, *Output* = 40

(C) *Input* = 5, *Output* = 45

(D) *Input* = 5, *Output* = 58

15. *You eat* $\frac{2}{6}$ *of a cake. How many equal parts does the whole cake have?*

Your Answer:

16. *Is* $\frac{1}{4}$ *closer to 0 or to 1 on a number line?*

(A) *Closer to 0*

(B) *Closer to 1*

(C) *Exactly in the middle*

(D) *At 1*

17. *To show* $\frac{5}{8}$ *on a number line, how many hops of* $\frac{1}{8}$ *do you take from 0?*

(A) 3

(B) 5

(C) 8

(D) 13

Find more at
ViewMath.com/SC-Grade3

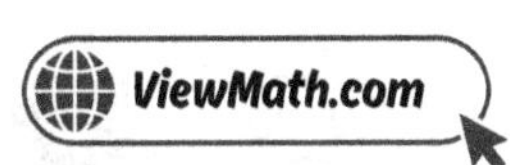

18. What whole number does $\frac{6}{3}$ equal?

(A) 1

(B) 2

(C) 3

(D) 6

19. When two fractions have the same numerator but different denominators, which fraction is greater?

(A) The one with the bigger denominator

(B) The one with the smaller denominator

(C) They are always equal

(D) You cannot compare them

20. A ribbon is $5\frac{1}{2}$ inches long. A piece of string is $4\frac{3}{4}$ inches long. Which is longer?

(A) The ribbon

(B) The string

(C) They are the same length

(D) Not enough information

21. Convert 3 kilograms to grams.

Your Answer:

22. What unit do we use to measure liquid volume?

(A) Grams

(B) Kilograms

(C) Liters

(D) Inches

23. How many cents are in $1.00?

(A) 1 cent

(B) 10 cents

(C) 50 cents

(D) 100 cents

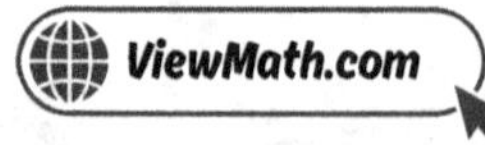

24. *What is $\$4.60 + \$3.85$?*

(A) $7.45

(B) $8.45

(C) $8.35

(D) $7.85

25. *A bar graph has a scale that counts by 10s. The bar for Red reaches the 3rd line. How many items does Red show?*

Your Answer:

26. *Which type of graph is BEST for showing measurement data like the lengths of worms in inches?*

(A) Picture graph

(B) Bar graph

(C) Line plot

(D) Tally chart

27. *Which shape is both a rectangle AND a rhombus?*

(A) Trapezoid

(B) Pentagon

(C) Square

(D) Triangle

28. *A rectangle is 10 feet long and 3 feet wide. What is its area?*

(A) 13 sq ft

(B) 26 sq ft

(C) 30 sq ft

(D) 33 sq ft

29. *Which shape has the greatest perimeter?*

(A) A square with sides of 5 cm

(B) A rectangle that is 8 cm × 3 cm

(C) A rectangle that is 6 cm × 4 cm

(D) A triangle with sides 7 cm, 7 cm, and 7 cm

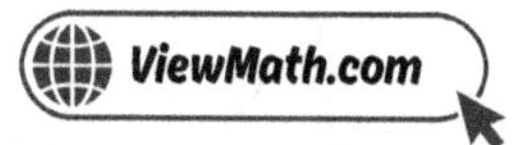

30. *A square is partitioned into 4 equal parts. All 4 parts are shaded. What fraction is shaded?*

(A) $\frac{1}{4}$                                                    (B) $\frac{3}{4}$

(C) $\frac{4}{4}$                                                    (D) $\frac{4}{1}$

Find more at
ViewMath.com/SC-Grade3

 # End of Practice Test 3 

Great job finishing the test!

 My Score

I got _____________ out of 30 questions right.

*Check your answers in the **Answer Key** at the back of the book.*

💡 *Review any questions you missed. That's how we learn!*

## 📊 Check Your Score Online!

Visit **ViewMath Academy** to enter your answers and see which topics you need to review. You can also explore lessons, take quizzes, track your scores, and save your progress!

**viewmath.com/score/3.1.SC.03**

*Or go to viewmath.com/score and enter code: 3.1.SC.03*

# Answer Key & Explanations

---

⭐ **Check Your Answers!** ⭐

First try each test on your own, then look here to check.

Read the explanations to learn from any mistakes ⭐

---

## ☑ Practice Test 1 — Answer Key

| | | | | | | | | |
|---|---|---|---|---|---|---|---|---|
| **1** B | **2** C | **3** C | **4** 377 | **5** B | **6** 800 | **7** $6 \times 8 = 48$ | **8** 45 | **9** 1 |
| **10** B | **11** C | **12** C | **13** 48 | **14** C | **15** $\frac{0}{3}$ | **16** B | **17** B | **18** $\frac{8}{8}$ | **19** A |
| **20** B | **21** C | **22** 3 L | **23** D | **24** B | **25** D | **26** C | **27** Rectangle | **28** D |
| **29** B | **30** B | | | | | | | |

---

💡 **Time to Learn!** 💡

Go through the explanations below, **especially for the questions you missed.**

Understanding why each answer is correct makes you a stronger math thinker!

👍 **Tip:** Circle any questions you got wrong, then read their explanation carefully.

---

## 📖 Practice Test 1 — Detailed Explanations

1  *In 7,431: the 7 is in the thousands place and the 4 is in the hundreds place.*

Find more at
ViewMath.com/SC-Grade3

2. $462$ has tens digit $6 \geq 5$, so it rounds up to 500. 428 and 449 round to 400. 551 rounds to 600.

3. Ones: $9 + 7 = 16$. Write 6, carry 1. Tens: $8 + 6 + 1 = 15$. Write 5, carry 1. Hundreds: $1 + 2 + 1 = 4$. The sum is 456.

4. Ones: $5 < 8$. Borrow: $15 - 8 = 7$. Tens: $3 < 6$. Borrow: $13 - 6 = 7$. Hundreds: $6 - 3 = 3$. The difference is 377.

5. Ones: $5 < 8$, borrow from tens. Tens is 0, borrow from hundreds. Hundreds is 0, borrow from thousands: $6 \rightarrow 5$, hundreds $\rightarrow 10 \rightarrow 9$, tens $\rightarrow 10 \rightarrow 9$, ones: $15 - 8 = 7$. Tens: $9 - 3 = 6$. Hundreds: $9 - 7 = 2$. Thousands: $5 - 2 = 3$. Difference: 3,267.

6. $248 \approx 200$, $175 \approx 200$, and $362 \approx 400$. So $200 + 200 + 400 = 800$.

7. 6 spiders with 8 legs each gives $6 \times 8 = 48$ legs.

8. $5 \times 9 = 45$. Skip count by 5: $5, 10, 15, 20, 25, 30, 35, 40, 45$.

9. Any number divided by itself equals 1.

10. The fact family for 9, 12, and 108 includes: $9 \times 12 = 108$, $12 \times 9 = 108$, $108 \div 9 = 12$, $108 \div 12 = 9$.

11. Step 1: $5 \times 8 = 40$ muffins. Step 2: $40 + 12 = 52$ muffins in all.

12. The rule is "divide by 3": $12 \div 3 = 4$, $18 \div 3 = 6$, $30 \div 3 = 10$. Missing: $? \div 3 = 8$, so $? = 24$

13. $(10 \times 4) + (2 \times 4) = 40 + 8 = 48$.

14. The rule is multiply by 6 ($3 \times 6 = 18$, $5 \times 6 = 30$). So $54 \div 6 = 9$.

**15**   0 parts shaded out of 3 total parts $= \frac{0}{3} = 0$.

**16**   $\frac{1}{2}$ is a bigger piece than $\frac{1}{6}$ because splitting into 2 parts gives bigger pieces than splitting into 6 parts.

**17**   $\frac{2}{3} + \frac{1}{3} = \frac{3}{3}$ (which equals 1 whole).

**18**   $1 = \frac{8}{8}$ because 8 eighths make 1 whole.

**19**   Same numerator (3 pieces). Fourths are bigger than eighths, so $\frac{3}{4} > \frac{3}{8}$.

**20**   $6\frac{3}{4} - 6\frac{1}{4} = \frac{3}{4} - \frac{1}{4} = \frac{2}{4} = \frac{1}{2}$ inch.

**21**   $3 + 5 = 8$ kg.

**22**   $4 - 1 = 3$ L.

**23**   A quarter is worth 25 cents.

**24**   $\$10.00 - \$4.50 = \$5.50$.

**25**   The scale counts by 5s, so the 4th line is $4 \times 5 = 20$.

**26**   Crayons shorter than 3 inches: 2 in. (2) $+ 2\frac{1}{2}$ in. (3) $= 5$ crayons.

**27**   It has 4 right angles and 2 pairs of equal sides (but not all 4 sides are equal), so it is a rectangle.

**28**   A square is a rectangle with all sides equal. Area $= 7 \times 7 = 49$ sq in.

(29) Lily multiplied $5 \times 3 = 15$, which gives the area, not the perimeter. The correct perimeter is $5 + 3 + 5 + 3 = 16$ cm.

(30) The $\frac{1}{4}$ slices are bigger because the pizza is cut into fewer pieces. More equal parts means smaller pieces: $\frac{1}{4} > \frac{1}{8}$.

## ✅ Practice Test 2 — Answer Key

| 1 | 3,705 | 2 | A | 3 | C | 4 | 357 | 5 | 2,654 | 6 | 600 | 7 | A | 8 | C | 9 | B |
|---|---|---|---|---|---|---|---|---|---|---|---|---|---|---|---|---|---|
| 10 | B | 11 | B | 12 | B | 13 | B | 14 | C | 15 | B | 16 | C | 17 | B | 18 | $\frac{32}{8}$ | 19 | > |

20. $6\frac{1}{2}$ inches  21. C  22. A  23. B  24. \$4.55  25. C  26. B  27. Hexagon

28. 81 sq in  29. B  30. 4

---

### 💡 Time to Learn! 💡

Go through the explanations below, **especially for the questions you missed**.

Understanding why each answer is correct makes you a stronger math thinker!

👍 **Tip:** Circle any questions you got wrong, then read their explanation carefully.

---

## 📖 Practice Test 2 — Detailed Explanations

(1) 3 thousands = 3,000, 7 hundreds = 700, 0 tens = 0, 5 ones = 5. The number is $3,000 + 700 + 5 = 3,705$.

(2) The ones digit is 8. Since $8 \geq 5$, we round up. 438 rounds to 440, not 430.

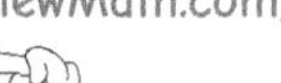

Find more at
ViewMath.com/SC-Grade3

3   Ones: $0 + 0 = 0$. Tens: $3 + 7 = 10$. Write 0, carry 1. Hundreds: $4 + 5 + 1 = 10$. The sum is $1{,}000$.

4   Ones: $5 < 8$. Borrow: $15 - 8 = 7$. Tens: $2 < 7$. Borrow: $12 - 7 = 5$. Hundreds: $5 - 2 = 3$. The farmer has 357 apples left.

5   Borrow across zeros: $5 \rightarrow 4$, hundreds $\rightarrow 10 \rightarrow 9$, tens $\rightarrow 10 \rightarrow 9$, ones $\rightarrow 10$. Ones: $10 - 6 = 4$. Tens: $9 - 4 = 5$. Hundreds: $9 - 3 = 6$. Thousands: $4 - 2 = 2$. Difference: $2{,}654$.

6   $358 \approx 400$ and $246 \approx 200$. So $400 + 200 = 600$.

7   $4 \times 3$ means count by 3s four times: $3, 6, 9, 12$. So $4 \times 3 = 12$.

8   $5 \times 6 = 30$. Skip counting by 5: $5, 10, 15, 20, 25, 30$.

9   20 split into 4 equal groups gives 5 in each group. $20 \div 4 = 5$.

10   $55 \div 11 = 5$. For repeating-digit dividends: $55 \div 11 = 5$.

11   Step 1: $3 \times \$7 = \$21$ spent. Step 2: $\$40 - \$21 = \$19$ left.

12   Multiples of 4: $4, 8, 12, 16, \ldots$ Multiples of 6: $6, 12, 18, \ldots$ The number 12 appears in both lists.

13   $11 \times 7 = 77$, so $77 \div 11 = 7$.

14   The rule is multiply by 8 ($2 \times 8 = 16$, $4 \times 8 = 32$). So $72 \div 8 = 9$.

15   Numerator is on top and denominator is on the bottom. So it is $\frac{2}{6}$.

16. A unit fraction always has 1 as the numerator. Examples: $\frac{1}{2}$, $\frac{1}{3}$, $\frac{1}{4}$, $\frac{1}{8}$.

17. $\frac{7}{8} = 7$ copies of $\frac{1}{8}$.

18. $4 \times 8 = 32$ eighths. So $4 = \frac{32}{8}$.

19. Same numerator. Fourths are bigger pieces than eighths. $\frac{1}{4} > \frac{1}{8}$.

20. The marker ends at the half-inch mark past 6, which is $6\frac{1}{2}$ inches.

21. $2$ kg $= 2{,}000$ g. Then $2{,}000 - 500 = 1{,}500$ g.

22. $20 - 13 = 7$ L.

23. Total: $\$0.85 + \$0.40 = \$1.25$. Change: $\$5.00 - \$1.25 = \$3.75$.

24. $\$12.00 - \$7.45 = \$4.55$. Borrow to subtract 45 cents from 0 cents.

25. $7 - 4 = 3$ fewer rainy days in February than April.

26. He measured 8 worms but only has 7 X marks, so he forgot to plot one measurement.

27. A polygon with 6 sides is called a hexagon. "Hex" means 6.

28. Area $= 9 \times 9 = 81$ sq in.

29. Add all side lengths: $4 + 6 + 8 = 18$ cm.

 Folding in half gives 2 parts. Folding in half again doubles it to 4 equal parts. Each part is $\frac{1}{4}$ of the whole.

## ☑ Practice Test 3 — Answer Key

| 1 B | 2 A | 3 C | 4 C | 5 A | 6 A | 7 C | 8 C | 9 C |

10  7. *Think:* $12 \times 7 = 84$, so $84 \div 12 = 7$.   | 11 C | 12 C | 13 C | 14 B | 15 6 | 16 A |

| 17 B | 18 B | 19 B | 20 A | 21  3,000 g | 22 C | 23 D | 24 B | 25 30 |

| 26 C | 27 C | 28 C | 29 B | 30 C |

---

### 💡 Time to Learn! 💡

Go through the explanations below, **especially for the questions you missed**.

Understanding why each answer is correct makes you a stronger math thinker!

👍 **Tip:** Circle any questions you got wrong, then read their explanation carefully.

---

## 📖 Practice Test 3 — Detailed Explanations

1. 9 *thousands* $= 9,000$, 0 *hundreds* $= 0$, 2 *tens* $= 20$, 5 *ones* $= 5$. So $9,000 + 20 + 5 = 9,025$.

2. 549 rounded to the nearest 10: ones digit is $9 \geq 5$, so 550. Rounded to the nearest 100: tens digit is $4 < 5$, so 500. $550 > 500$.

3. In the standard algorithm, when a column totals 10 or more, you write the ones digit and carry (regroup) the tens digit to the next column on the left.

4. When the top digit is smaller, you borrow (regroup) 1 from the next column to the left. This adds 10 to the current column.

5. Borrow across zeros: $7 \to 6$, hundreds $\to 10 \to 9$, tens $\to 10 \to 9$, ones $\to 10$. Ones: $10 - 4 = 6$. Tens: $9 - 5 = 4$. Hundreds: $9 - 2 = 7$. Thousands: $6 - 3 = 3$. Difference: 3,746.

6. Nearest 10: $460 + 220 = 680$. Nearest 100: $500 + 200 = 700$. The exact answer is 682, so rounding to the nearest 10 gives a closer estimate.

7. 5 groups of 6 means $5 \times 6 = 30$ stars.

8. Multiplying by 2 is the same as doubling. $9 + 9 = 18$.

9. Any number divided by 1 equals itself. $9 \div 1 = 9$.

10. Division is backward multiplication. Ask: $12 \times\ ? = 84$. Since $12 \times 7 = 84$, the answer is 7.

11. Step 1: $36 \div 4 = 9$ cookies per plate. Step 2: $9 + 3 = 12$ cookies per plate.

12. The pattern adds 7 each time (multiples of 7): $28 + 7 = 35$.

13. $11 \times 12 = (11 \times 10) + (11 \times 2) = 110 + 22 = 132$.

14. $5 \times 8 = 40$. The correct output for input 5 is 40.

15. The denominator tells the total equal parts. The cake is cut into 6 equal parts.

16. $\frac{1}{4}$ is only 1 hop from 0 out of 4 hops total, so it is much closer to 0.

**17** $\frac{5}{8}$ means 5 copies of $\frac{1}{8}$, so take 5 hops from 0.

**18** $\frac{3}{3} = 1$, so $\frac{6}{3} = 2$.

**19** Same numerator means the same number of pieces. A smaller denominator means bigger pieces, so the fraction with the smaller denominator is greater.

**20** $5\frac{1}{2} = 5\frac{2}{4}$, which is greater than $4\frac{3}{4}$. The ribbon is longer.

**21** $3 \times 1,000 = 3,000$ g.

**22** We measure liquid volume in liters (L).

**23** $\$1.00 = 100$ cents.

**24** Cents: $60 + 85 = 145$ cents $= 1$ dollar and 45 cents. Dollars: $4 + 3 + 1 = 8$. Answer: $\$8.45$.

**25** The scale counts by 10s, so the 3rd line is $3 \times 10 = 30$ items.

**26** Line plots are best for measurement data. They show each individual measurement as an X on a number line, making it easy to see how values are spread out.

**27** A square has 4 right angles (like a rectangle) AND 4 equal sides (like a rhombus). So a square is both a rectangle and a rhombus.

**28** Area $= 10 \times 3 = 30$ sq ft.

**29** Square: $4 \times 5 = 20$ cm. Rectangle $8 \times 3$: $(2 \times 8) + (2 \times 3) = 22$ cm. Rectangle $6 \times 4$: $(2 \times 6) + (2 \times 4) = 20$ cm. Triangle: $7 + 7 + 7 = 21$ cm. The $8 \times 3$ rectangle has the greatest perimeter of 22 cm.

Find more at
ViewMath.com/SC-Grade3

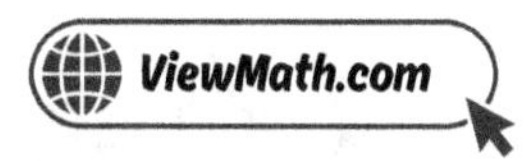

 All 4 out of 4 parts are shaded, so $\frac{4}{4}$ is shaded. $\frac{4}{4} = 1$ whole.

## Great job checking your work!

Keep practicing and you'll be a math star!

Find more at
ViewMath.com/SC-Grade3

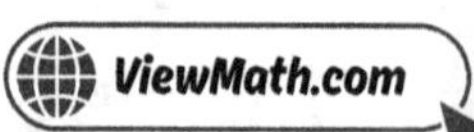